Acid Rain

Stephen Sterling

Wayland

Our Green World

Acid Rain
Atmosphere
Rainforests
Recycling

LG/5574

C6041677 99

Cover: Power stations produce gases that cause acid rain.
These trees have probably died from the effects of acid rain.

Series editor: Philippa Smith
Series designer: Malcolm Walker

First published in 1991 by
Wayland (Publishers) Ltd
61 Western Road, Hove
East Sussex BN3 1JD, England

British Library Cataloguing in Publication Data
Sterling, Stephen
  Acid rain.
  1. Acid rain
  I.Title II. Series
  551.5771

  ISBN 0-7502-0137-1

Typeset by Kudos Editorial and Design Services, Sussex, England
Printed in Italy by G.Canale & C.S.p.A., Turin
Bound in France by AGM

▲ *Canadian maple trees are being damaged by acid rain.*

Forest damage

'My family has lived on this farm for over 100 years. Many of the old trees are now damaged by acid rain and are dying from insect attacks. A couple of years ago a strong wind blew some of them down. The younger trees will never reach this age.'

Ingemar Zachrisson, Sweden

If you look at old buildings, you might notice that the stone is dissolving. This happens naturally, but acid rain can speed up the process.

The Statue of Liberty in New York had to be restored because of acid rain damage. ▼

▲ Acid rain has damaged this English church.

▲ *Here in Germany, traffic is being diverted to avoid smog – a mixture of pollution and fog.*

In some places, even breathing can be dangerous for people with health problems like asthma. In Germany, pollution in the air is sometimes so bad that people are advised not to go out.

More and more people are becoming worried about acid rain. Because of this, action is being taken to tackle the problem. For example, some countries now have strict pollution controls.

Join the club!

In 1984, ten countries in Europe formed a '30 per cent club'. They promised to cut down the amount of sulphur dioxide they produced by at least 30 per cent (one-third) by the year 1993. Although Britain did not join the club, it was agreed that sulphur dioxide would be cut by nearly two-thirds by 2003.

◄ *A layer of pollution drifts over a valley in Germany. The Germans quickly took action when they found that pollution was harming their forests.*

▲ *A Swedish lake is treated with lime.*

One way of fighting the effects of acid rain is by dropping lime into lakes. This is called 'liming'. Lime is an alkali made by crushing limestone rock. When it is put into lakes, the acidity of the water is reduced. Sometimes lime is spread over ice-covered lakes. When the ice melts, the lime falls into the water.

If the pH level can be raised to pH 6.5, plants and animals can return to the lake. But liming has to be repeated every 2 to 5 years and it is expensive.

These dying trees might be helped if lime were put on the soil, but it is difficult to spread lime in thick forests.

Sometimes, lime is spread on the ground and this helps the soil and plants. But liming will not solve the problem of acid rain. We must cut down on the amount of sulphur dioxide and nitrogen oxides that are produced.

Power stations provide the electricity we need for making industrial products, heating, lighting, cooking and transport. But power stations also cause some of the acid rain. If we used electricity more carefully, we would not have to make so much, so there would be less pollution.

▼ *Power stations produce electricity, but they also produce the gases that cause acid rain.*

Much of the world's energy comes from coal, oil and gas. But electricity can also be made in ways that do not cause acid rain.

◄ *This power station in France makes electricity from the heat of the sun.*

These windmills in the USA produce electricity. ▼

▲ *This is a hydro-electric power station. It makes electricity from falling water.*

Another source of energy that does not produce sulphur dioxide pollution is nuclear power. But people worry about the dangerous **nuclear waste** that is produced.

▲ *Some power stations now have special cleaning devices to remove most of their sulphur dioxide pollution.*

Another way to reduce pollution is to take the sulphur out of coal before it is burnt. Or sulphur can be taken out of the fumes by fitting cleaning devices called **scrubbers** to power station chimneys.

When a tree is cut down for timber, it is best to take only the trunk. The roots and branches can be left to rot. This helps stop acidity building up in the soil.

▼ *This forester is cutting down a tree for timber.*

Can you think of ways in which we can all help cut down air pollution? Write them down and then look at the list on page 39.

◄ *If more people used buses and trains instead of cars, there would be less air pollution. Why is this?*

Cars are a major cause of air pollution. Using lead-free petrol is a good idea, but harmful gases are still given off from car exhausts. Most of these gases can be reduced by fitting the exhaust with a special filter. It is called a **catalytic converter**.

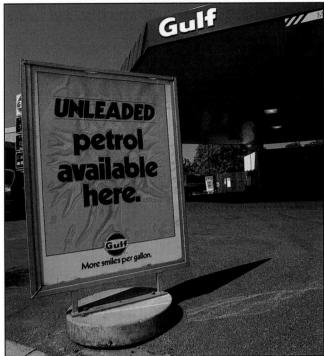

▲ *Using lead-free petrol is one way cars can make less air pollution.*

◄ *The exhaust fumes from vehicles contain nitrogen oxides, sulphur dioxide and other harmful gases.*

In the USA and Japan, laws have been passed to reduce pollution from cars. Similar laws will soon be passed in European countries.

It costs a lot of money to clean up and prevent acid rain. But money must be spent in order to stop even more damage to the environment. Some countries have begun to take action, and the amount of sulphur dioxide produced in Europe has fallen.

Some lakes in Scandinavia are recovering a little, but air pollution needs to be cut still more.

Protesters climb a factory chimney to draw attention to the pollution it produces. ▼

The European parliament often meets to talk about environmental matters. ▶

Saving a Swedish lake

'Our lake has been saved by schoolchildren from the city. Many sorts of wildlife were disappearing because the lake was becoming acidic, and the children wanted to save it. They helped us spread lime on the ice. The next summer they measured the acidity of the water. It had fallen from pH 5.7 to pH 6.8.'

Bjorn Hansen

As well as cutting sulphur dioxide, we must also cut the amount of nitrogen oxides that we produce. About half the nitrogen oxides in air come from car exhausts.

The USA has strict laws controlling pollution from cars. Other countries have made slower progress, but all new cars in the **European Community** countries will have to be fitted with a catalytic converter by 1993.

▼ *Los Angeles in the USA is having to reduce air pollution from cars. Can you see the smog in this picture?*

Reducing pollution

Here are several ways in which we can help to cut down air pollution:

Put on extra clothes when it is cold, instead of turning up the heating. This will mean that less fuel is burnt.

Turn off the lights when they are not needed, so that power stations do not have to produce so much electricity.

Ask the drivers in your family to drive cars more slowly so that the engine produces less pollution. Drivers could also turn the engine off when a car is in a long queue or waiting at traffic lights.

Use buses and trains instead of cars – they can carry far more people in one trip. This cuts down the amount of pollution. Better still, walk or cycle whenever you can.

Can you think of any other ways?

Acid rain is a big problem that is damaging our environment. But not everyone agrees about all the causes and effects of acid rain.

People often disagree about other issues affecting the environment too. It is up to us to make up our own minds, but first we must learn about the issues and listen to different points of view.

▼ *We need to learn about our environment.*

This sign in Tokyo, Japan, shows levels of air pollution.

We can all do something to help the environment. Think what you could do about this, and ask your teacher for some more ideas.

▼ *These healthy trees and lake are signs of a clean environment. Let's try to keep it healthy!*

Action at home

You can test to see if the rain where you live is acid.

Cut the top off a plastic drinks bottle, and put a plastic bag inside the bottom half as a lining.

Place the bottle outside, away from buildings and trees, and support it by tying it to a stick pushed into the ground.

When it has rained, lift out the plastic bag, and test the water with universal indicator paper. Use the colour scale on page 7 to see how acid your rainwater is. Take the sample in a screwtop jar for testing at school if necessary.

Repeat this test several times, but each time use a fresh plastic bag.

Natural rainwater has a pH of 5.6. If the pH levels of your samples are more acid than this, you have acid rain.

Each time you do the test, you should keep a note of what direction the wind comes from when it rains. If the water is more acid when the wind is from a certain direction, there may be a reason for this that you can work out. For example, has the wind come from an area with power stations or much traffic?

If the rain is acid where you live, what do you think could be done to make it less acid?

Glossary

Acid A substance which has a pH of less than 7.0. The opposite to alkali. Acidic is the word to describe something that is acid. Acid foods like lemons have a sharp or sour taste.

Alkali A substance which has a pH of more than 7.0. The opposite of an acid. Alkaline is the word to describe something that is alkali.

Breed To produce babies.

Catalytic converter A filter fitted to car exhausts to help remove pollution.

Coniferous trees Trees with cones and needle-like leaves (such as pine or fir).

Environment A plant or animal's surroundings, including air, water, soil and other plants and animals.

European Community (EC) A group of European countries which work together.

Fossil fuels Sources of energy such as coal, oil and natural gas which have been formed over thousands of years from the remains of dead animals and plants.

Minerals Substances formed naturally in rocks and the earth, such as coal and tin.

Nitrogen oxides Polluting gases formed from nitrogen in the air. They are produced when fossil fuels are burnt.

Nuclear waste Very dangerous waste material from the nuclear power industry.

Nutrients Substances that are nourishing and help a plant or animal to grow.

pH scale pH is the measure of acidity. The pH scale goes from 1 to 14 (pH1 is the most acid). There is a 10 times difference between each pH value, so pH5 is 10 times more acid than pH6.

Pollution Harmful substances in the environment.

Scandinavia The countries of Sweden, Norway and Denmark in northern Europe.

Scrubber A device attached to a factory or power station chimney to clean pollution, such as sulphur dioxide, from the fumes.

Sulphur dioxide A polluting gas formed from sulphur. It is produced when fossil fuels are burnt.

Timber Wood which is suitable for making buildings and furniture.

Picture acknowledgements
The publishers would like to thank the following for allowing their photographs to be reproduced in this book: John Baines 23 below, 28; Bruce Coleman Limited 4 (Wayne Lankinen), 5 above (John Shaw), 5 inset (Kim Taylor), 6 (John Brownlie), 8 above (Steve Kaufman), 11 (George McCarthy), 12 (Colin Molyneux), 15 (P A Hinchcliffe), 18 (John Shaw), 21 (Hans Reinhard), 23 above, 24 above (Adrian Davies), 24 below (Norman Tomalin), 25 (N Schwartz), 34 (Norman Owen), 40 (Colin Molyneux); CEGB/DIPA Photo Library 32; Greenpeace 36; ICCE 22 (Rod Redknapp), 35 above (Mark Boulton); Andre Maslennikov/IBL Sweden 16, 27; Oxford Scientific Films cover (Ronald Toms), 14 (G A MacClean), 29 (Steve Littlewood); Rex Features 37 (Boccon); Science Photo Library 30 above (Gazuit); Topham Picture Library 41; ZEFA 8 below, 10 (Hunter), 17 (Adam), 19, 20 (Mosler), 26 (M Becker), 30 below (T Braise), 31 (Bramaz), 33 (Hunter), 35 below (H Schmid), 38 (G Juckes), 42. The illustrations are by Jane Pickering 18, 22; Malcolm Walker 7; and Brian Watson 6, 9, 13, 15, 21.

Finding out more

Useful addresses

Acid Rain Foundation
1630 Blackhawk Hills
St Paul MN 55122, USA

Acid Rain Information Centre
Dept of Environment and
    Geography
Manchester Polytechnic
Chester Street
Manchester M1 5GD

Campaign for Lead-Free Air
    (CLEAR)
3 Endsleigh Street
London WC1H 0DD

Council for Environmental
    Education
University of Reading
London Road
Reading RG1 5AQ

Friends of the Earth (UK)
26-28 Underwood Street
London N1 7JQ

Friends of the Earth (Australia)
National Liaison Office
366 Smith Street
Collingwood
Victoria 3065

Friends of the Earth (Canada)
Suite 53
54 Queen Street
Ottawa KP5CS

Friends of the Earth (New
    Zealand)
Negal House
Courthouse Lane
PO Box 39/065
Auckland West

Greenpeace (UK)
30-31 Islington Green
London N1 8XE

National Society for Clean Air
136 North Street
Brighton
East Sussex BN1 1RG

Watch
22 The Green
Nettleham
Lincoln LN2 2NR

World Wide Fund for Nature
    (UK)
Panda House
Weyside Park
Godalming
Surrey GU7 1XR

Books to read

Acid Rain by Tony Hare (Franklin Watts, 1990)
Acid Rain by John Baines (Wayland, 1989)
Acid Rain by John Baines (wallchart and notes - Pictorial Charts Educational Trust, 1989)
Atmosphere by John Baines (Wayland, 1991)
The Blue Peter Green Book by Lewis Bronze, Nick Heathcote and Peter Brown (BBC Books, 1990)
The Green Detective up the Chimney by John Baines (Wayland, 1991)
The Young Green Consumer Guide by John Elkington and Julia Hailes (Gollancz, 1990)

Index